Unexplained

Unsolved,

Unsealed

Mysteries of World War 1

(Volume 5)

Gerald Burns

Copyright © 2021 – Gerald Burns

All rights reserved.

No part of this publication may be reproduced, stored in a retrieval system, or transmitted, in any form or by any means, electronic, mechanical, photocopying, recording or otherwise, without the prior written permission of the author

See other books in the series

## Lonely Hearts
Jilted

Kiss, the Stockpiler, goes to war.

In search of petrol

The Hungarian vampire

In search of a killer at war

## The Floating Coal Mine
The USS Cyclops

The Disappearance

The Search and Last Sightings

The Eccentric Captain

## Lieutenant Richter
The Misfortunes of the Ub-65

Lieutenant Richter's Death

The Horror Begins

Lieutenant Richter succeeds

Discovery of the wreckage

## Celtic Mystery
The Attack

The Aftermath of the Attack

Theories

## Mutiny Mystery

Trouble is brewing

Full-Scale Mutiny

# Introduction

Lonely Hearts

A Hungarian man goes to fight for his country during WW1. In his absence, the police discover a shocking secret in his home: metal barrels filled with pickled human remains. A vampire is on the battlefield, and nobody is safe.

The Floating Coal Mine

The last message received from its crew said, "Weather Fair, All Well." The ship and its 309 passengers were never seen again. 103 years later, it is all questions and more questions. Not a single discovery.

Lieutenant Richter

A submarine commander dies from an explosion. But crew members and sailors of the U-boat keep seeing him around the dark passages of the U-boat. The ghost issues warnings to beloved colleagues before he does the inevitable and takes them with him

Celtic Mystery

Fifty-nine soldiers apparently go missing during a battle with enemy forces. No traces of their bodies, ammunition, or clothes are ever found. Were they captured by the enemy, or did something more supernatural happen in the woods?

Mutiny Mystery

A riot breaks out in a soldiers' camp. Five people are killed, and 23 are seriously injured. To date, historians still have no clear picture of what happened or who ordered the shooting. Justice continues to hang in the air, unserved.

# Lonely Hearts

In the early 1900s, a young man moved into Cinkota, a small town near Budapest, the capital of Hungary. His name was Bela Kiss, and he was a successful tinsmith that neighbors would grow to love and admire in no time. Kiss was born in Izsak, Hungary, to his father, Janos Kiss, and his mother, Verona Varga. Years after discovering his atrocities, there were several reports about an incestual relationship with his mother.

For one thing, Bela Kiss was known to be a man who fancied self-development and the acquisition of knowledge. Even with his poor educational background, Bela developed himself so much that he could hold his own against the great scholars and academicians that lived in Cinkota. He was also known to dabble in occult practices. He has been described by people who knew him as a voracious reader who would delve into any topic and gather as much information as possible until he understood it for himself. To say he was an inspiring young man would be an understatement.

Kiss was married twice, but he moved into Cinkota with his wife, Marie, a lady who was about ten years his junior. Dissatisfied with her marriage to an older man, Marie would go around town searching for younger men to love. During this time, she met a man called Paul Bikari, and they started a not-so-discreet romantic affair. The neighbors got a hold of this information, and it didn't take long for their tongues to start waggling. Soon enough, Kiss found out that his wife was cheating on him with a younger man.

## Jilted

A few weeks later, in December of 1912, Kiss started to tell everyone who cared to listen that his young wife had eloped with her younger lover. He gained the pity of all and sundry as he was a man everyone loved. To comfort himself and reduce the loneliness he now experienced, the jilted lover decided to hire a woman (Mrs. Jakubec) to serve as a housekeeper in his home. Mrs. Jakubec tended to Kiss' needs like she would her son and the two came to love and respect immensely. She was allowed to go anywhere around the house except Kiss' study, where he spent most of his time each night.

The jilted man started inviting different good-looking and well-dressed ladies to his home to comfort his broken heart further. These women ranged from wealthy widows, ladies he'd met in the local bar to prostitutes from brothels where he was a regular visitor. Around the same time, a man simply known as 'Hoffman' began placing ads in the 'Lonely Hearts Section' of the city's newspaper. He presented him as a great fortune teller and a matrimonial agent. The direction to his office led straight to Kiss' home. Widows and young ladies who were desperate to wed flooded his apartment in the hope of finding the right man to marry.

Most of the women found him appealing, tall, and charming in a lovely way. His fame spread further around town. He proposed marriage to some of them for a hefty fee, and once they agreed and paid the money, he would allow them to move into his home and share his bed. But the neighbors and the new housekeeper noticed a weird trend. Many of the women who went into Kiss' apartment were never seen walking out. Surprisingly, no one suspected him enough to raise the alarm.

## Kiss, the Stockpiler, goes to war.

While Kiss acted as a fortune teller, his neighbors noticed that metal drums were steadily delivered to his property. To quench their curiosity, Kiss assured his neighbors that he was stockpiling petrol in case of a war outbreak in Europe. Everyone at that time knew that the world was on the verge of another war, and governments worldwide were preparing. Kiss assured his friends that his stored commodity would be equally rationed when the time came, and this would greatly help the people to get through a hard time of the war.

A few weeks later, detectives from Budapest arrived at Cinkota with inquiries about two widows who had left their home with the intention of visiting a certain Hoffman. None of their families and friends had heard from them since their journey down to the Cinkota. The people around knew that Kiss had been taking in women, but they didn't know that he had used a pseudonym to conceal his identity. The detectives found nobody in Cinkota by the name of Hoffman, so they ruled the case inconclusive and determined that he too had vanished into thin air. War was brewing, and the police and military had more important dealings on their hands than to focus on some missing widows, so the case went cold in no time.

1914 was the year World War I started in earnest. That year, Kiss was drafted into the army, after which he was handed his uniform and rifle and asked to join the ranks at the battlefield. Before leaving, he called his housekeeper and asked her to keep the charge of his property until his return. With that, he turned his back and went to fight for his country.

# In search of petrol

In 1916, the Budapest police arrived at Cinkota armed with information about a house filled with stockpiled petrol. On arrival, they were pointed to Kiss' home where they met Mrs. Jakubec. At first, Mrs. Jakubec denied them entry into the house. And she had a valid point. Kiss had entrusted the safety of his house into her hands so it would make no sense to just give it up to these strange men at her door. It took a while for one of the police men to convince her to let them into the house to see the petrol for themselves. They told her they would buy it from her and save the money until Kiss' return. Reluctantly, the housekeeper let them into the building and took them down to the compartment where Kiss stored his metal barrels.

Down in the cellar filled with drums of varying sizes, one of the policemen took a nail and punctured one side of a barrel to determine its content. A liquid with a rotten stench leaked out to the floor, and an unpleasant smell filled the air. It was evident that the barrel contained neither diesel nor petrol. This was undoubtedly the odor of something dead. The men went ahead and pried the top of the barrel open for further investigation. In it, they found the naked body of a woman submerged in alcohol, her head alone above the liquid. In shock, one of the policemen sent word to Detective Chief Karoly Nagy and requested his presence.

The detective and his team set to work on arrival, opening all the other drums to determine their contents. They found seven additional barrels, which all contained naked women strangled to death.

Detective Karoly ordered a search of the wooded area around Kiss' home. There, they found more disturbing discoveries of victims who'd been buried in the ground, many of them preserved in alcohol beforehand. Their search ended with the recovering of 17 other bodies in the surrounding area, bringing the total number of victims to 24 (plus the 7 found inside the house). Among these remains were the bodies of Kiss' wife and her young lover, Paul Bikari.

## The Hungarian vampire

Detective Karoly knew that it wouldn't take too long for the news to hit the airwaves and reach Budapest. His first action was to send a letter to the military leader on the battlefield, informing him about Kiss' misdeeds. Kiss was to be taken into custody immediately.

Due to the nature of the killings, Detective Karoly also believed that Kiss had acted with an accomplice, so a warning was sent out to all postal services in the area to withhold any messages addressed to Bela Kiss. This, he believed, would make it almost impossible for the accomplice to warn Bela of his impending doom.

Seeking more information about the murder, Detective Karoly also ordered Mrs. Jakobec's arrest. The distraught housekeeper adamantly held on to claims that she knew nothing about Kiss' murderous nature. She told police that she believed the drums had been used to store up liquor for the entertainment of Kiss' numerous lady friends. Plus, he'd also warned her never to go close to any of them. After a series of interrogations, it was determined that she knew nothing about the murders, so she was released. But that was not before she offered to help out in her own small way.

Mrs. Jakobec agreed to show investigators some hidden rooms in Kiss' house and also provide them with keys to gain entrance and search them. At first, she took them to his master's bedroom where they found nothing of importance to their investigation. Mrs. Jakobec then pointed to another door in one corner of the room that had been concealed by drapes. She pulled out the keys in her possession and opened up the door. Inside, investigators found a windowless, dark room with a leather chair and desk coated in dust. The bookshelf provided enough evidence of Kiss' wrongdoings. They contained plenty of books that provided instructions on strangling a victim and the proper procedure for mummification accurately.

They found a leather bag underneath the table that bore evidence of correspondence between Bela Kiss and numerous women searching for husbands. He had many pictures of them stacked up in his bag. The earliest interaction with these ladies dated back to 1903. It also became clear that Kiss was the s0-called Mr. Hoffman, putting ads in the Lonely-Hearts section of newspapers. His documents showed that he'd received marriage proposals from about 174 women, and he'd accepted 74 of those.

From the letters, detectives noticed a trend. Kiss never went forward with any of the ladies who had relatives nearby or people who would notice their disappearance. Whenever he met such ladies, he would leave them, never to return. His documents contained old court documents filed by two ladies who claimed he'd defrauded them without fulfilling his side of his deal. The court closed the case without question after both women disappeared mysteriously.

One of his victims went by the name Katherine Varga. Varga was a widow with a successful clothing business that she'd sold before moving over to Cinkota to marry her new husband. Another victim was Margaret Toth, a lady from a family with means. Margaret's mother had paid a considerable dowry to her daughter's lover before allowing her to move into his house. Once there, Kiss convinced the lady to write a letter to her mother, informing her of her intention to move to America and start a new life. As soon as she was done with the letter, Kiss strangled her and mailed the letter to her parents.

For those he was interested in, Kiss would invite them to his house, where he would rob and kill them before he pickled their remains in alcohol and sealed them in his metal barrels. Many of the victims' bodies showed puncture marks from which he had drained blood from inside them. This led many to believe that Kiss practiced vampirism in which he would consume the blood of his victims to satisfaction.

## In search of a killer at war

In October of 1916, detective Nagy received information of Kiss' whereabouts. The letter stated that the killer was lying sick in a hospital bed in Serbia. Nagy immediately picked up his handcuffs and started his journey to Serbia. Unknown to the detective, the disappointment that awaited him was one that would stay with him for years. He reached the place too late, for he found another dead soldier on the bed where Kiss was supposed to be recuperating. Kiss had somehow gotten legit information about Nagy's arrival, and he'd placed the corpse of a dead soldier on the bed before escaping unnoticed. Detective Nagy returned home with the biggest disappointment of his career, one that would hunt him for the rest of his life.

There were numerous sightings of Bela Kiss by people in different parts of the world. One claimed that he made it to Romania and started a small business before being caught and imprisoned for burglary. A soldier who'd participated in World War I reported on a fellow legionnaire who referred to himself as 'Hoffman' and strongly resembled Bela Kiss. The said 'Hoffman' escaped the area before the Hungary police could reach him.

In 1932, another report came of Bela Kiss sighting in Manhattan, New York. A detective Henry Oswald claimed that he'd seen Bela Kiss roaming among the multitude in Times Square. It is believed that he worked as a janitor somewhere in the city. Unfortunately, the janitor in question fled before he could be questioned.

At this point, you would agree that it is almost impossible to find out what really happened to the serial killer in this mystery.

Did he die on the battlefield or move to another part of the world to continue his murderous acts? We may never know for sure. One certain thing is that this doctor of lonely hearts lived life on his own terms till his demise. Justice never caught up with him.

# The Floating Coal Mine

The oceans of the world contain mysteries in various forms. From stories about missing ships to sailors stranded on uncharted islands down to tales of wild nautical beasts that torment pirates and their loots. We remain in awe of these narratives and continue to imagine all the surprises that await us out there at sea. One of these terrifying mysteries is the one that involves the disappearance of the USS Cyclops.

## The USS Cyclops

The USS Cyclops was one of the four proteus-class colliers designed for the US Navy in anticipation of World War I. The battleship was named after Cyclops, one of the giants heavily featured in Greek mythology. The ship was only the second to bear that name, and its design and construction were considered a breakthrough in navy battleship technology at the time. This makes its disappearance one of the most enduring mysteries of World War I.

The USS Cyclops was completed and launched on the 7th of May, 1910. Constructed in a Philadelphian site, the steel-hulled ship was considered one of the largest and fastest US Navy ships of its time. It was about 540 feet long, 65 feet with a steam 15 knots, larger than a standard football field. The Cyclops could haul more than 12000 tons of coal for its energy requirement throughout its journeys. Newspapers of the time described it as a floating coal mine on sea. In fact, the ship was primarily designed to help in the refueling of the US Navy fleet. This meant that its occupants faced a big risk on an outing because it was prone to catching fire with that great deal of coal sitting in its cargo.

In early 1910, the USS Cyclops operated in the Baltic Sea and the American East Coast. The ship was employed during the operation of Vera Cruz to assist in the transportation of refugees. When America declared war on Germany in April 1917, the Cyclops was commissioned entirely for transportation and was reinforced with 50-caliber guns and medical supplies to be transported from John Hopkins Hospital to Saint-Nazaire Hospital, France. At this point, it became a valuable asset for the US Navy, used to transport coal supplies worldwide to battleships in need. This designation took Cyclops on assignments as far away as Brazil down to Nova Scotia.

# The Disappearance

A few months later, Cyclops arrived in Brazil to load manganese ore for transportation to Baltimore, America. The ship's crew was unfamiliar with handling this product, which was much denser and heavier than the normal coal. They ended up filling their vessel with 10,000 tons of the product when it could only carry 8000 tons. Just before departure from Brazil, the ship's captain filed a report about a cracked cylinder somewhere in the vessel. A quick survey showed that this was true, but mechanics also confirmed that the cargo was totally fine and ready to sail. Bear in mind that the Cyclops was an 8-year-old ship with a faulty engine set to transverse the rough seas of the Atlantic Ocean. This is without mentioning the threats of war. Plus, its route would take the ship and its crew through the Bermuda Triangle.

The vessel left the shores of Brazil and traveled down to Barbados, where it restocked for the rest of the journey. It left Barbados on the 4th of March, 1918, for a trip that should take only nine days to get to America. Along the way, something terrible happened to the ship, something we can only guess. Nine days later, Cyclops was not seen in Baltimore. The last message received from its crew said, "Weather Fair, All Well." The vessel was never seen again.

# The Search and Last Sightings

Soon enough, a search for the missing ship started around the area where it was last seen along its initial route. One sailor and his crew claimed they'd seen it sailing off towards Virginia, but there was no success finding the vessel along that route. Naval vessels searched isolated beaches and scanned major and minor trade routes for debris or anything that could point to a disaster. Their crew sent radio signals every day with the hope of receiving information from the Collier. They found nothing and received no reply. By July 1918, Franklin D. Roosevelt, the Assistant Navy secretary at the time, announced that the ship and all its 309 inhabitants were presumed lost. The tragedy instantly became the largest loss of life at sea that was unrelated to navy combat.

Reports were received that the collier had been captured by German raiders or sunk by U-boats, but none of these had any evidence to back them. The cyclops would have been a target for the Germans and a huge victory after a takedown. After weeks of investigation, it became increasingly clear that the Germans had not been in that area at all. There was no trace of debris which is the norm after a foundering at sea.

Another claim was put forward that the vessel had met its end in the hands of bad weather and rough seas. That was a strong possibility, but then why didn't they send out a distress message if such a thing had happened? Plus, there were no storms reported around the area when the ship disappeared.

Some have speculated that the Cyclops was sucked down the gigantic drain of the Bermuda Triangle of North America, just like many other oceanic vessels before it. One source in a popular magazine summarized that a giant octopus had risen from the ocean and wrapped its mighty tentacles around the collier before pulling down into the depths of the ocean.

One Marvin Barrash, a descendant of one of the firefighters aboard the vessel, has his theory. After decades spent researching and collecting relevant information, Barrash came up with the idea that a series of major mechanical failures caused the ship to lose balance before a minor wave tipped it over. He thinks this may have happened somewhere around the Puerto Rico Trench, where the deepest part of the Atlantic Ocean is located.

## The Eccentric Captain

In the absence of a clear explanation for Cyclops' disappearance, we look towards the issue of the eccentric captain of the collier, Captain George W. Worley. His is one of the most intriguing theories surrounding the ship's loss. The commander was born in Germany – where he was known as Johan Fredrick Wichmann – and he'd illegally migrated into the US, where he changed his name and took on a new identity.

Not only was he an eccentric leader who adopted a slew of strange authoritative doctrines, but Captain George was also terribly hated by his crew members. This hatred sprung from his manner of approach and the harshness with which he dished out instruction and meted out his punishments to the offending staff of the vessel. His unusual nature involved inspecting the vessel's compartments in nothing but a hat, a walking stick, and his underpants. So, what we have here is the case of an American naval vessel captained by a person of German extraction in a war against German naval vessels. If that isn't a perfect recipe for mutiny, then I don't know what is. In the face of accusations from crew members of the ship, the captain would flare up and sometimes order their arrest or execution. All of this was just to hide his true identity. Higher naval officers also stood behind him and cleared him of all charges before reinstating him into his position, where he sought revenge against all those who planned to expose him.

Some believed that this man might have handed over the ship to his German brothers in exchange for some glorious reward.

\*\*\*

Many still express hope that Colossus' remains would be found someday, and this mystery would have a befitting explanation. Given the advent of modern undersea exploration technology, their hopes could be rewarded in the future. Underwater explorers continue to research and dive deeper into the ocean in search of the treasures of the USS Cyclops, but years go by without any tangible discovery. Years ago, excitement grew when a diver discovered a shipwreck down in the ocean just off Florida's coast. In the end, it was just a German fuel ship from World War II. As of 2018, Baltimore Republican Rep. Andy Harris had plans to erect the first monument regarding the lost ship and its inhabitants.

# Lieutenant Richter

In retrospect, one can simply refer to World War I as that war that saw the Allied Forces and Central Powers challenge each other by introducing updated war technology and strategic approaches and battle tactics. Both parties tasked their greatest military minds to come up with innovations that would help them gain the upper hand and win the war with little stress.

On the one hand, Britain was renowned for its naval prowess at the time, having some of the largest sea vessels in the 1900s. The only problem they faced was that many of their vessels were already on the verge of falling apart. Most of them were old and refurbished cargos from the last 1800s. On the other hand, Germany was well aware of the flaw and was determined to use it to their advantage. In truth, they didn't have the naval strength to face Britain out there at sea, so they decided to take the battle down into the water by investing heavily in the production of powerful undersea boats, more popularly known as submarines or U-boats. The Germans employed these U-boats as weapons against naval fleets and for enforcing a blockade against their enemies.

In the heat of World War 1, Germany possessed about 30 of these U-boats, incomparable to the over 60 owned by the

British navy. But the Britons never saw a need to use U-boats for their battles on the water. They considered it cowardice – the act of diving underwater and attacking the enemy from beneath. Unfortunately, there is no fair game in war, and the German continued to improve their submarine technology and successfully take down many British vessels. Over and over, the Germans proved to be an unstoppable force undersea as they successfully launched a series of attacks and completely decimated many warships.

For example, on the 22nd of September, 1914, the German U9 attacked and sank three British Cruisers. Over 1000 sailors and crew members were buried at the bottom of the sea that day. Months later, a U-20 attacked and sank another cruise liner traveling towards the Irish coast. 1190 lives were lost that day alone. In all of this, the Royal Navy could only stand and watch in horror as it lost many of its vessels without the capacity to protect itself or fight back.

## The Misfortunes of the Ub-65

The German Ub-65 was built at the Vulcan Shipyards in Hamburg in 1916 by the Imperial Navy to decimate more enemy ships. But even before the completion of its construction and the launching of the submarine, its sailors and commanders already had unsettling premonitions about this small sea vessel. She was considered unlucky before being allowed to prove herself.

The misfortunes of this boat started during her construction, for there was an unusual number of accidents that occurred within this period that gave cause for alarm. The first instance happened when the hull was being laid. A huge steel girder that was being lifted fell off when its chain snapped midair. The falling girder instantly killed one of the men who had been standing just underneath it while the other remained trapped under heavy metal, wailing in agony as his comrades tried to free him. His torture continued for two hours until he finally gave up the ghost. Later investigation showed that the chain used in lifting the girder was in perfect condition. There was no explanation whatsoever for its sudden snap.

On another occasion, crew members became trapped in an engine room when leakage occurred, and dangerous diesel fumes permeated the air around them and choked their lungs. Later that day, their colleagues noted three men were unaccounted for. They were all found dead in the engine room where the deadly fumes had asphyxiated them. Again, there was no explanation for the leakage. In fact, the system had been a newly installed one.

This long line of strange and spooky events did not stop after the submarine's deadly construction. Its death count continued and manifested itself again the day a test run was carried out to determine the UB-65's preparedness for undersea performance. During the trial, the submarine encountered a strange storm that came out of nowhere, and huge waves crashed into one side of the U-boat, sweeping one of the bystanding crewmen into the ocean, never to be seen again.

During another test run to ascertain the U-boat diving ability, the crew experienced another set of technical difficulties that resulted in a fracture of one of the ballast tanks. Water surged into the newly-formed holes around the boat and weighed it down closer to the ocean's bed. The crew panicked, and for over 12 hours, they battled with various equipment with the hope of taking the submarine back to the surface of the ocean. All of this happened as their oxygen supply reduced and they suffocated. In the light of a miracle, the crew finally found a way to work her up to the ocean's surface before any lives were lost.

One would think the last experience would mean an end for the UB- 65. But, no! The Germans were unrelenting in their struggle to gain the upper hand in the war. This time they were rewarded with a gas leak that killed another two men. At this point, tongues began waggling with superstitious beliefs about the boat's disaster. Determined to quench these beliefs and cheer up his men, the captain of the vessel found ways to take minds off the numerous deaths of recent times and turn their focus to their importance in the ongoing war. He succeeded for a while until Lieutenant Richter's death.

## Lieutenant Richter's Death

Despite the unfortunate events attached to the Ub-65, the submarine prepared for its official maiden voyage in 1917. The day before takeoff, a small group of men were loading in the torpedoes when one suddenly detonated without any warning. The resulting explosion somewhat damaged one part of the submarine and injured five sailors. The second officer in command at the time (Lieutenant Richter) was killed instantly during the blast. After weeks of repairs and refurbishments, UB-65 was set to make her way into the North Sea, where she would combat the Royal Navy. By this time, new members had joined its crew to replace those lost in the past weeks. What the crew members did know was a journey full of horror awaited them at sea.

# The Horror Begins

In October 1917, UB-65 embarked on its first mission, which turned out to be a great success for the crew members. During that mission alone, the U-boat successfully sank five enemy vessels, the most notable being the British Corvette HMS Arbutus. These strings of success stories caused the team to briefly forget the ominous air hanging around their boat as they celebrated and planned more attacks.

Trouble struck one evening when the commander of the submarine, Captain Martin Schelle, had one of the crew members barge into his room while he snacked to inform him of a terrifying situation. Schelle dumped his food on the bed and sprung out to see what it was. He met one of his sailors shivering on the floor, muttering out some words in fear. The sailor explained that he'd gone into the submarine's conning tower to take one last glance of the horizon when he noticed a man in a coat standing just a few feet away from the bow. The sailor claimed that he'd called out to the man to identify himself, and the figure turned to face him with a weird smile on his face. It turned out to be Lieutenant Richter, the commander who'd died just months ago.

Captain Schelle was incensed with the story he'd just heard. He flew into a rage, calling the shocked sailor all sorts of names, and blamed his hallucinations on drunkenness and deep-seated fear. The Captain then warned him not to spread rumors around the boat regarding ghost sightings before leaving the terrified men and walking back to his compartment. But at this point, it was already too late. News of the incident had traveled throughout the small boat.

Calmness returned to the submarine for a while until it was punctured again by a scream coming from the boat's control room. This time a crew member had gone there to carry out some minor assessments when he felt someone tap him from behind. He'd turned back to see Richter, who mouthed an inaudible warning before vanishing into thin air. While backing away from the control room, the crewman lost his balance and fell backward down the ladder that led to the room. He broke his leg on landing. The first crew members who arrived at the scene to attend to him also claimed that they saw Richter peering down at them from the hatching in the control room before he vanished one last time.

From then onwards, the sighting of Lieutenant Richter's ghost became the order of the day in the submarine. He was once seen in the dark toilet of the submarine, looking through the windows one dark stormy night. The dark alleyways that lead to rooms became dreaded passages as many crew members claimed they'd seen shadows bearing Richter's form slip through walls. Captain Schelle's threats did nothing to stop the spread of these tales. One crew member and his colleague claimed they'd been working on a piece of equipment when they noticed Richter standing just by their side before the ghost looked away and walked through a bulkhead.

The crew members and their submarines had many failed outings during this period as they could not take down any enemy vessels. Schelle had every reason to believe this was happening because of the dead morale of his teammates. None of them were willing to work in isolated places alone, making it even more challenging to assign jobs. The captain quickly started making arrangements to replace some crew members hoping that new intakes would mean the end of these absurd stories about the dead commander.

The incident that broke the camel's back happened one day when a torpedo man inspecting a piece of equipment in the bow compartment saw Richter walk past him before vanishing through a steel wall. Propelled by a feeling of apprehension and maddened by the never-ending tales of ghosts, the crewman made his way through the boat's passages and hurled himself into the North Sea. Captain Schelle and his comrades could only stand on the deck and watch this man as he screamed and swam away from the haunted U-boat, never to be seen again.

# Lieutenant Richter succeeds

By now, the story about the haunted UB-65 was a pain to the German naval force. A clergyman was invited to perform an exorcism on the boat and pray for some of its crew. After that, sailors who wished to be replaced were quietly exchanged with those from other U-boats.

The submarine was again released with its crew member for another round of patrols. They completed two missions successfully, without any hitch or ghost sightings. Little did they know that Richter had other plans.

On July 31st 1918, UB-65 met her tragic end. An American submarine and its crew would be the only witness to this final fate. Lieutenant Augustine Grant and his crew had come across the German vessel just off the Cornish Coast. They noticed the submarine was already tilting to one side even as it floated on the water. With his binoculars, Captain Grant noticed a man in a dark coat and a hat standing on the deck, undisturbed by the situation as he stared out into the sea. As his men loaded their torpedoes to fire down the enemy's vessel, they all witnessed the U-boat shake before an explosion erupted from the water as the boat sank and buried itself and all its sailors. The American captain and his crew hurriedly swan towards the wreckage, searching for survivors, but that search would end up futile. It remained that way for 85 years.

## Discovery of the wreckage

Despite the initial search for the U-boat, the wreckage of the haunted submarine was not found until 2004 when an underwater crew shooting for a Channel 4 TV series finally saw an unidentified U-boat lying deep down at sea. Its identity was confirmed to be the infamous UB-65 after a U-boat historian, Dr. Axel Niestle, was called in to take a look at the wreckage. His investigation revealed that there hadn't been a weapon attack on the boat, a revelation that conflicted the official German record that had it that its own torpedoes had damaged the boat. The open hatches of the submarines suggested that some of its inhabitants had tried escaping for their lives without any success. To date, U-boat experts have not been able to identify the leading cause of the boat's disastrous fate. All of these later discoveries only deepen the mystery surrounding the UB-65 boat.

It is evident that none of the sailors fabricated their stories since such lies could lead to extreme punishments from their commandants and ridicule from their colleagues. The fact that almost all of the inhabitants of this boat had their own creepy experience with Lieutenant Richter's ghost goes to say that something mysterious really happened on that vessel. If so, then why was the Second Commander's ghost so adamant about taking down his former crew members with him? Is it possible that we do not know the full story about his death? The questions are seemingly endless.

The incident leading to the foundering of UB-65 has also been blamed on mass hysteria. Given the circumstance in which the men found themselves, living in such a confined vessel where they had to spend most of their time underwater uncertain of the next attack or instruction from their commander, could it be that they succumbed to the pressure of mass hysteria as it spread its tentacles throughout the ranks? Sailors, being who they are and given their line of work, are mostly superstitious men and women. It is possible that they so bought into the reality of their illusion until the monster they feared started to take a form for itself. That, too, is just a theory about the incident. All we know for sure is that something sinister and unpleasant happened the day UB-65 fell and dragged all those men to their death.

# Celtic Mystery

The Celtic mystery refers to the disappearance of some members of the 10th Battalion of the Australian first division, a battalion known for its fierceness in battle and its many conquests. For their fierceness, the Battalion was popularly referred to as the 'Terrible 10th' given their long history of bravery in the face of danger. Two of its members once earned the revered Victoria Crosses, the highest honor bestowed on any soldier fighting for the British army. It was this Battalion and its commander that defended the ANZAC Cove on the Gallipoli Peninsula of Turkey.

After another successful outing by troops of the Allied Forces in the Battle of Broodseinde, Field Marshal Douglas Haig insisted that the dreaded German forces had been beaten down and were on the verge of collapse. There was reliable information that the Germans were considering a withdrawal from the Belgian coast. For this reason, Douglas and his men planned an offensive attack on the Germans to push them back further and capture the Passchendaele Ridge. This land tussle would become the Battle of Passchendaele.

The following week, while they hatched their plans, light but consistent rain began to fall. This new condition made advancement hard for members of the Allied Forces due to the water-logging of the area. On the other hand, the German forces found this condition favorable to their cause as it provided them the opportunity to retreat into the steadier ground. The British army struggled to advance because many of the planned routes had become too muddy to navigate at this point. Taking unplanned routes meant that they could become exposed and end up as easy targets for the German forces. The commanders of the Allied Forces decided to come up with a more complex strategy to maintain their control of the battle despite the weather condition.

The resulting plan was straightforward and relatively simple: The terrible 10th Battalion would launch a diversionary attack on the Germans at Celtic Wood. They would attack at dawn and blow up the German dugouts before retreating after the release of a flare signal. At the sight of the flare, another Australian Battalion to the north would launch a larger attack to protect the British advance on its northern flank. But then, the Allied Forces knew that they would have to successfully fool the Germans into believing that the attack from the 10th Battalion was the main attack, not just one meant to distract them from the real thing.

The first strategy for deception would be to attack at dawn since diversionary attacks are known to occur at night. The second strategy involved replacing the normal box barrage used to protect raids with the rolling barrage used in full-scale attacks. The members of the 10th Battalion agreed to the plan since they knew that this tactic would greatly favor their troops if it turned out successful.

## The Attack

The 10th Battalion launched their attack at 5:20 Am on the 9th of October. The attacking men consisted of 78 soldiers plus seven officers (t0tal of 85 men) on their way to Celtic Wood. They were led by 22-year-old Lieutenant Frank Scott. The commander was a great leader of his battalion, and his wisdom proved vital on the battlefield that day. Once the German launched a counterattack to defend themselves, vocal commands passed from leaders to soldiers became almost impossible to receive due to the eardrum-wrecking sounds around the woods. The leaders of the 10th Battalion soon discovered that the Germans also outnumbered them, but they pushed forward and matched to execute a successful frontal attack.

What followed after that was hand-t0-hand combat between both forces with heavy casualties recorded on the enemy's side. The Allied Forces had the battle in their hands, and they were unwilling to let go of their dominance. But something happened that turned the table. The German forces laid down fire between the Scott-led Battalion and the Australian trenches, eliminating the possibility of a successful retreat. While the Australians struggled in the hopeless situation, the German gained reinforcement. It didn't take long for them to gain the upper hand in the battle and start gunning down the Australians left, right, and center. Eye witness accounts claimed that Sergeant William Cole managed to fire the flare before the enemy gunned him down.

The remaining Australian soldiers on the battlefield found it very hard to make their retreat back to the safety of their camp. Seeing their comrades gunned down mercilessly, many of them decided to play dead by lying in the mud with the hope of successfully escaping in the darkness of the night.

## The Aftermath of the Attack

The day after the dreaded battle, Battalion Commander Lieutenant Colonel Maurice Wilder-Neligan sent a message home that said, "I am only able to account for 14 unwounded members of the party". One of the survivors claimed that only ten of his comrades had made it back to the Australian lines, while another said he'd counted 14 soldiers. After a comprehensive count, the 14 soldiers were found and gathered in the Australian camp. Seven others later returned with terrible wounds, and five bodies were eventually found in the woods after the battle. This left 59 soldiers unaccounted for. Among these 59, only 22 of them had known names. It seemed that the other 37 had been unidentified figures recruited into the armed force. The question now is: "What happened to the 59 soldiers, and what happened to the record of the unnamed 37.

## Theories

Many explanations have been put forward to explain the whereabouts of those bodies. One of these theorized that the bodies might have sunk deep into the mud, never to be seen again. But this theory is quite problematic because bones have a way of returning to the surface after some time. It is no news that bones of slain soldiers are still being discovered all around Europe to date.

Professional historians have gone to the Celtic Wood area to ask the nearby inhabitants if they had any memory of human remains lurking around. The portion that was Celtic Wood has now been plowed and turned into a vast field. None of the workers ever saw any bone lying around as they worked on the area.

There is a belief that the German soldiers captured the surviving members of the Battalion as prisoners before massacring them and burying them someplace else. The issue with this theory is that the names of these captured soldiers never turned up on any prisoners' lists exchanged through the Red Cross. During World War One, German troops were not so popular for the massacre of their captives. Also, the British barrage on the battlefield would make it almost impossible for the Germans to walk away with 59 Australian soldiers. Strangely enough, there is no record of the battle in German military records, which suggests that something questionable happened that day, something the Germans never want to face.

As always, people have come to theorize that something supernatural happened right there at Celtic Wood. There are claims that the soldiers had simply turned into the woods before they vanished into thin air. Others have blamed the disappearance on alien abduction.

Two Australian historians – Chris Henschke and Robert Kearney – who studied the Celtic Mystery for years came up with their theory. They believe that the heavy fire engaged by the opposing forces met the soldiers in the middle, killing them and pulverizing their bodies. This could be the reason why their remains were never found since they'd been reduced to bits of blood and bones scattered around the muddy area. This theory rings true since the phenomenon was an everyday occurrence during World War One. Many of the soldiers sent to war were unaccounted for and simply filed as 'missing,' presumably captured by enemies.

# Mutiny Mystery

At the end of World War 1, with the Armistice signed on the 11th of November, 1918, many soldiers who participated in the war began to anticipate their return home to family and loved ones. The action had died down, and battlefields had dried up. The Canadian servicemen were moved to Kinmel Park in North Wales, where they were told they would have to remain temporarily until their return home.

Kinmel Park had been in existence since before the war. It was only converted to a camp for soldiers awaiting repatriation after the war in 1918. The camp comprised 20 smaller camps, all with their accommodation huts, mess halls, and training grounds. One military hospital was stationed at one side of the camp. Outside the gate was a mini-market consisting of stores, bars, and restaurants to cater to the troops.

The soldiers' living conditions were poor, with as many as 15,000 men sharing the camp. The food was even more terrible, with little or no coal to warm the men who mostly slept on the cold floor. Europe was experiencing severe winter at the time, but as many as 40 men were forced to share a room for 30 as they took turns to sleep on the floor throughout the night. Not only were the men homesick, but the Spanish Flu was raging in other parts of Europe and moving closer to their camp. More than 80 soldiers had met their death at the hand of the flu in other camps around Kinmel. Also, the pay was slow to come by, meaning the men had to find other means to survive before the next money entered their hands from above. They were all eager to return home fast, fully aware that the first to reach the Canadian shores would be the first to get good jobs there.

Canada also had its own problems. For example, the country only had two ports through which ships could enter the country. The other ones were either too small or had been covered by ice, making sea transportation even harder. This meant that they would need at least 18 months to get everyone back home.

In good faith, these servicemen accepted all of these conditions, their minds comforted with the hope of getting home soon. After all, the torture would only last for a while before they would be shipped back home, back to the comfort of their beds. Or so they thought.

The problem was that they didn't have the ship to take the men home. The only one available was the Northland which was rusted both inside and outside and had been declared unsafe to carry such a large number of men over the Atlantic. The ship was literally falling apart, and Canadians at home would not dare to transport their comrades in such a monstrous atrocity. So, the men at Kinmel Park were asked to be patient and wait for the arrival of a better vessel to take them home. Getting access to the next available ship meant filling out 30 documents and answering over 300 questions, which the men found unnecessary and boring. They were paraded in the camp twice a day and made to undergo training and exercises, even though the war was over.

## Trouble is brewing

Resentment and angst started building at Kinmel park when some of the soldiers learned that functional sea vessels were being sent to pick up American troops to take them home while ignoring their plight at Kinmel. Many of the US soldiers had experienced little or no battle at the fields, which made it even more annoying. The American troops hadn't even been away from home as long as the Canadians at Kinmel.

By the 4th of March, 1919, almost five months after the armistice, the bulletin boards at Kinmel showed that no ship had been sent to pick up the men. They were growing restless and rebellious. That morning, an officer tried to order some of the men out to march, but no one moved as much as a foot.

Seeing the rising tension in the camp, Colonel Colquhoun sent out a letter to London, pleading for an urgent intervention to get the troops back to Canada. The letter's messenger (Colonel Thackery) was ordered to stay in London until the ship was assigned. Thackery soon received credible information that the ships would be made available to the troops on the 15th of March. For some reason, Thackery hesitated in sending out the telegram to Kinmel and succeeded in doing so hours later. But the time it reached the camp, it was too late. The riot was on.

By the night of the 4th of March, tensions had already reached boiling point. Some had broken into the canteen and stolen all the valuables, knocking off owners who tried to get in their way. To the soldiers, this was their way of getting back at the shop owners, who were notorious for overcharging them for goods because the inhabitants of Kinmel park had nowhere else to purchase their necessities. That night the soldiers destroyed the market and took all the valuables they could lay their hands on.

## Full-Scale Mutiny

A message was immediately sent to the quarter Colonel Colquhoun shared with other top-ranking officers. He asked his men to lock up all the other untouched stores and empty every keg of beer before meeting the protesters to ask them to calm down. They gently shoved him away and continued their looting until the following day. By the time the sun was up, the devastation had become alarming and Colquhoun had to order all ammunition to be gathered and locked in a bunker to prevent the looters from shooting each other. Some of the rioters were also taken prisoners that evening.

The following day, the soldiers gathered in front of the guard housing, demanding the release of their comrades. One lieutenant colonel armed his men and charged them to hold back the men seeking the freedom of their friends. It was at this point that one of the protesters was bayoneted. The lieutenant had armed his men even though the protesters were unarmed and only seeking justice.

Later that afternoon, some incensed soldiers got hold of rifles and returned to the guardhouse to exchange bullets with the lieutenant's men. That gunfight resulted in the loss of five men and injury of between 20 to 30 soldiers, with 78 arrested for participating. Among those arrested, 28 were given prison sentences ranging from 3 months to 10 years imprisonment, although many of these were later reduced to 6 months only. There were no charges regarding the loss of lives, and all of the inquiries into the case failed to bring forth any conclusive finding regarding the person responsible for those deaths.

There is a belief that more people were killed at that riot than the officially provided number of 5. A local coroner, Fredrick Llewellyn Jones, convened a failed inquest into the killing that received no cooperation from the Canadian Army. Most of the witnesses among the protesters had been hurriedly shipped home after the mutiny for obvious reasons, so the coroner only had the officers directly involved in the killing to talk to, and they made it a hell show for him until he gave up in frustration and the case went cold. The subject simply became one shrouded in military secrecy. More than 100 years later, the families of the slain men only know that their relatives died in the hands of their comrades and not from shots fired from opposing forces on a battlefield during the war. Their death could have been prevented if the correction had been taken.

Private Joseph Young was one of the men who died of a bayonet wound to his head. The inscription on his grave reads, 'Sometime, sometime, we'll understand.' In response to questionnaires received from historians seeking the truth, one of the veterans wrote back, "It wouldn't have happened if the officers had only treated us like men."

Other books in the series

**Let us hear your thoughts**

If you enjoyed this book, please support Gerald Burns by going over to Amazon to drop a short and honest review. It would be fully appreciated. Also help us in spreading the word in any way possible by getting more people to read this book. It would mean the world to us.

Thank you very much!

Gerald Burns' obsession with true crime stories and the dark psychology of serial killers started on a flight to flight to New York in 2004 while reading a book on the Manson murders. Few months later he read a story about a young waitress who had gone missing after her shift, never to be seen again. She simply vanished.

Since then he became interested in true crimes stories, stories of disappearances and the weird histories of the world. He started writing out about some of these disappearances in magazines and had people send in their suggestions and thoughts about the cases. The interest for this led him to start writing out some of these stories for the enjoyment of other true crime addicts.

You can connect with him on Twitter, Instagram and Facebook.

Printed in Great Britain
by Amazon